Underwater Musicians

by D. M. Souza

Carolrhoda Books, Inc./Minneapolis

The publisher wishes to thank Andrew Bass, Associate Professor of Neurobiology and Behavior at Cornell University, for his help in the preparation of this book.

Carolrhoda Books, Inc. c/o The Lerner Publishing Group
241 First Avenue North, Minneapolis, MN 55401 U.S.A.

Website address: www.lernerbooks.com

Library of Congress Cataloging-in-Publication Data

Souza, D. M. (Dorothy M.)
    Underwater musicians / by D. M. Souza ; photographs by ENP Images
       p.   cm. – (Creatures all around us)
    Includes index
    Summary: Describes a variety of sea creatures, including fishes, crustaceans, and walruses, and how and why they make the sounds they do.
    ISBN 1-57505-097-8 (alk. paper)
    1. Fishes—Behavior—Juvenile literature. 2. Marine animals—Behavior—Juvenile literature. 3. Sound production by animals—Juvenile literature. 4. Fish sounds—Juvenile literature.
[1. Fishes. 2. Marine animals. 3. Animal sounds.] I. Title.
II. Series: Souza, D. M. (Dorothy M.). Creatures all around us.
QL639.3.S69 1998
597.159'4—dc21   96-52996

Manufactured in the United States of America
1 2 3 4 5 6 – JR – 03 02 01 00 99 98

Underwater Musicians

The beluga whale's song is like the chirping and squeaking of birds.

The underwater world is filled with a chorus of sounds. Drumbeats and grunts, wheezes and sighs, whistles and clicks move through the waters. The sounds come from many different directions and fill the water with a strange kind of music. Who are the mysterious performers that make these sounds, and what are they doing?

They are shrimp, lobsters, odd-looking fish, seals, walruses, dolphins, and giant whales. All these sea creatures use sounds to send messages in a shadowy world that can be as dark as the night.

Some of these creatures moan when their young wander out of sight. Others hum or whistle to attract mates. They grunt at neighbors that are bothering them or make popping sounds to frighten away enemies. Sound is as important to these sea animals as sight is to an eagle. But what is sound, and how does it travel?

sound
waves

THWACK
ACK ACK
ACK

Place a ruler on a table so that part of it hangs over the edge. With one hand, hold the ruler firmly on the tabletop. Snap the other end several times with your thumb. Watch the ruler vibrate, or move quickly back and forth, and listen to the sound it makes. Every sound begins with a vibration.

The air around us is made up of millions of invisible particles, called **molecules.** When we speak, shout, clap our hands, or snap a ruler, these molecules begin to vibrate.

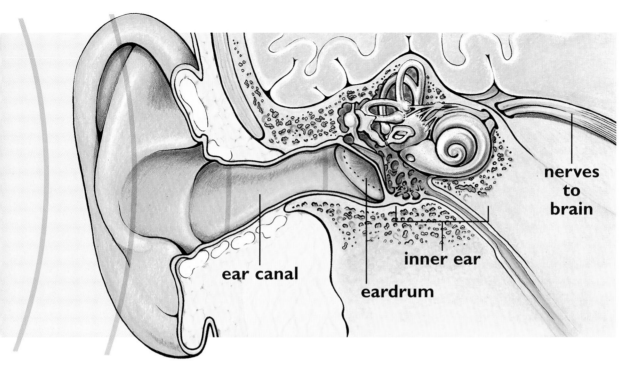

nerves
to
brain

ear canal

inner ear

eardrum

Have you ever tossed a rock into a pool of water and watched the waves spread out in all directions? In a similar way, vibrating molecules, or sound waves, spread out in different directions. Some of these sound waves reach your ear, vibrate down an air-filled tube, and strike a thin layer of skin in your ear, called an **eardrum.** This eardrum vibrates and makes parts of your inner ear move. Then your nerves, thin threads that act like telephone lines, send signals to your brain. Suddenly you know what sounds you are hearing.

Many sea creatures hear the way we do. Fish, walruses, and most seals don't have outer ears, but they have inner ones just below the surface of their skin. They also have nerves like ours that send signals to their brains. Whales and dolphins not only hear through their inner ears, they also pick up vibrations through their lower jaws. These vibrations travel to their ears and then to their brains.

Sounds play an important part in the lives of many underwater animals. As different sea creatures sing their songs or play their instruments, others answer. Let's follow a few of these performers and discover how and why they make their music.

Some whales and dolphins, like these Atlantic spotted dolphins, hear songs of their relatives through their lower jaws.

Drummers and Hummers

Drums, members of the croaker family, get their name from the sounds they make.

Shiny, blue-gray fish swim above the sandy bottom of the ocean. They purr, whistle, creak, and croak. Farther away, some of their relatives hum and drum. All these fish belong to a large family of music-makers, the croakers.

Like many fish, the croaker has a **swim bladder,** or air-filled organ, that keeps it from sinking. By changing the amount of air in its bladder, the fish can move up or down, or float in the same place. But the croaker and other similar fish also use this swim bladder as a musical instrument.

Several strong muscles surround the swim bladder. When a fish vibrates these muscles, purring, whistling, humming, or drumming sounds move through the water. The swim bladder acts like the sound box of a guitar. In a guitar, the hollow space inside the instrument makes the sounds of the strings louder. A swim bladder works the same way: it turns up the volume of the vibrating muscles.

The northern sea robin, a type of croaker, moves up and down by changing the amount of air in its bladder.

Croakers often gather in schools when it's time to mate.

During mating season, when animals look for a partner to produce young, male and female croakers often gather in **schools,** or groups, of up to a million fish. The first thing they do is sing to one another. In some **species** (SPEE-sheez), or kinds, of croakers only the males sing. In others, both males and females serenade one another. Some croakers start singing softly in the evening, get very loud around midnight, and then gradually quiet down. Others make their sounds only in the evening or early morning. Croakers are one of the noisiest kind of fish in the sea. Some croakers are so loud, people on boats can hear croakers that are far beneath the surface of the ocean.

Like croakers, toadfish use their muscles and swim bladders as sound-makers. They sing to win mates and scare away **predators** (PREH-duh-turz), the creatures that like to eat them. One toadfish, the midshipman fish, has been nicknamed the "canary bird fish" because of its lively songs.

During spring, midshipman fish move to shallow waters along the western coast of the United States. Here, they prepare to mate and search for places around rocks to hide their eggs. The males hum night and day to attract females. Their songs may last from several minutes to an hour and then begin all over again.

Silently the females come closer to the music-makers. After they lay their eggs in nests, they leave the area. But the males stay and guard the nests until the young hatch. When predators come near, the males grind their teeth together and grunt loudly.

Once the young midshipman fish hatch, the males return to deeper waters where they search for food, grunt at enemies, and wait for next spring's mating concert.

Grinders and Poppers

A rover grouper makes its home near coral reefs.

Have you ever been frightened and whistled a happy tune to calm yourself? Groupers seem to do this all the time. When frightened, these members of the sea bass family send out a chorus of moos, grunts, thumps, and rumbles. They make these sounds by grinding their teeth and by vibrating muscles against their swim bladders.

The garibaldi, or ocean goldfish, which lives off the southern California coast, becomes nervous when other animals enter its territory. It swims around grinding its **pharyngeal** (fuh-rin-JEE-ul) **teeth,** the teeth lining its throat. The clicking, scraping sounds warn strangers to stay away.

The garibaldi grinds its teeth to keep other animals out of its territory.

The spiny lobster rattles its antennas if an octopus or other enemy comes too close.

Fish aren't the only sea creatures that make sounds when they're afraid. Spiny lobsters, which belong to a group of animals called **crustaceans** (kruh-STAY-shunz), have two sets of antennas on their heads. When an octopus or other enemy threatens the lobster, it rattles its small antennas and warns the predator not to come any closer. The lobster's warning call can be heard from 150 feet away.

16

The pistol shrimp, another crustacean, lives on the sandy bottom of the ocean. It has a large claw and a smaller, thumblike one. If disturbed, it rubs the two claws together and *pop!* It's as if someone has pulled the trigger of a toy pistol. The shrimp uses its "pistol" to scare away would-be attackers. The popping vibrates the water with such force that it often knocks out worms and other prey the shrimp likes to eat.

Some shrimp, like the pistol shrimp, make snapping, popping noises by rubbing their claws together.

A dumeril filefish makes weird sounds with its teeth.

Some fish make music when they're having dinner. Filefish, for example, have two sets of teeth. One set, the **incisors,** sticks out of their tube-shaped mouths. When a filefish finds something to nibble on, it rubs these teeth together. Suddenly, a noise that sounds like fingernails scratching a chalkboard travels through the water. Another set of teeth, the pharyngeals, makes a grinding, crunching noise. All this rasping, grinding, and crunching may sound weird to us. But to a filefish that has just found a delicious meal, it's music.

Graybar grunts, swimming in a school, send messages to one another by grinding their teeth.

Grunts use sounds to keep in touch with one another as they swim in and around coral reefs. These fish get their name from the gruntlike sounds they make when grinding their pharyngeal teeth. Grunts are true noisemakers. Even if you lift a grunt out of the water, it may keep on grunting.

Some **mollusks** (MAH-lusks), which are soft-bodied animals without backbones, have shells that make sounds when they move. Blue mussels, for example, live off the eastern coast of the United States. They anchor themselves to rocks, to the hulls of ships, or to each other, so they aren't tossed around in rough seas. The mussels hold themselves in place with strong lines that they produce within their bodies. When the mussels want to move, they simply break the lines. This makes a snapping, popping noise that lets others know they're leaving.

Thin threads that anchor blue mussels to rocks pop when the animals move from place to place.

Walruses haul themselves out of the Arctic Ocean and rest on blocks of ice.

Knockers and Chuggers

In the Far North, a herd of about 20 female walruses, known as cows, and their calves haul themselves out of the icy waters of the Arctic Ocean. These animals, which are **mammals** like us, crowd together on a rocky shore. If there isn't enough space, they pile on top of one another. Their grunts and barks rise above the creaking sounds of the moving ice.

Meanwhile, a 3,000-pound bull, or male walrus, swims nearby. For the next few weeks, the bull will try to coax the cows to mate with him. When the cows enter the water to feed, he dives with them. While they rest on land or on sheets of floating ice, he stays underwater and sings to them.

His song begins with vibrations that sound like knuckles hitting a wooden door. These knocks are followed by tapping and bell-like sounds. After about 7 minutes, the bull comes up for air. He floats on his stomach and makes several more knocking sounds between breaths. Minutes later, he dives and continues his underwater concert. The whole performance may last more than 24 hours and can be heard up to 10 miles away.

A Weddell seal can make a variety of sounds.

Elsewhere, in the frozen waters around Antarctica, relatives of the walrus also sing strange love songs. All seals make a variety of sounds, but Weddell seals are probably the most musically talented.

Each spring, females return to their favorite places near shore. They flop onto moving masses of ice and give birth to pups that have been growing inside their bodies since the previous spring or summer.

Like walruses, male Weddell seals set up underwater territories nearby and wait for the females to enter the water. When they do, the males' songs, filled with knocks, whistles, pops, and cracks, can be heard even through thick sheets of ice. Males also defend their territories by screaming "Goh, goh, goh." Both males and females make a series of chugging and chirping sounds when they're disturbed.

Pups stay on the ice and often imitate the music of the adults. If they are hungry or frightened, they bawl, "Aa, aa, aa." Their mothers immediately recognize their voices, and call back or swim toward them.

A Weddell seal pup waits on the ice while its mother takes a swim.

Walruses and seals both have vocal chords in their throats, like we do. When the animals make sounds above water, these chords vibrate the air and others hear them. But as they dive deep underwater, their mouths are closed. How do they sing their songs underwater?

Some scientists believe that males make their knocking music by snapping their tongues against the roofs of their mouths. The movement of their jaws makes the water around them vibrate. Their bell-like music may come from two pockets in their necks. Females do not have these pockets, but males can inflate their pockets until they are as big as basketballs. When they let the air out, it sounds like ringing bells.

A walrus surfaces near some rocks.

Atlantic spotted dolphins hunt together for food.

Clickers and Whistlers

On a clear summer morning off the coast of Argentina, a **pod,** or group, of close to 50 dusky dolphins moves toward deeper water. These mammals swim side by side, about 30 feet apart, as if they are part of a giant underwater parade. Suddenly the dolphins dive. Within seconds they circle a school of anchovies, small fish that are a special treat.

The dolphins swim around the fish in a tighter and tighter ring. Gradually they herd the fish toward the surface, where they begin feasting on them.

27

Dolphins, like this common dolphin, make a high-pitched clicking noise when they force air through their blowholes.

How did the dolphins know where to find the anchovies and when to dive? First, they sent out high-pitched clicking sounds. Scientists believe dolphins make these sounds by forcing air through one or two **blowholes,** or nostrils, on their heads. Next, the clicks echoed off the school of fish, and vibrated the water back toward the dolphins. The animals picked up the vibrations through thin bones in their lower jaws. From here, signals raced to their brains and let the dolphins know how far away the anchovies were. This method of using sound and echoes to find prey is known as **echolocation.** Dolphins and some whales use it when they feed in dark waters.

Dolphins travel long distances and hunt in large groups. As many as 1,000 dolphins can be in one pod. Imagine what all those clicks sound like!

In the waters off the coast of Washington and British Columbia, a different kind of dolphin makes its home. It is known as the killer whale, but it isn't a whale—it's a dolphin. Whistles, high-pitched squeals, and screams vibrate the waters. Scientists are not certain how killer whales make these sounds. But they do know that they sing to keep in touch with one another. When the pod is ready to move farther out to sea, the whistles and squeals become louder and last for several minutes. It's as if the message is being passed from one killer whale to the next.

Killer whales whistle, squeal, and scream when they meet one another.

A pod of killer whales hunt for fish close to shore.

When two or more pods of killer whales meet after being separated for some time, they sound like a roomful of noisy relatives. Whistles and squeals are loud and lively and can be heard more than 5 miles away. Young killer whales squeak, squawk, and whistle as they playfully race, chase, and bump into one another.

Singers

Of all the creatures in the sea, whales are probably the best-known musicians. The unusual music of these mammals has even been captured on CDs and cassettes for people to hear and enjoy.

The beluga whale, for example, chirps, squeals, toots and screams like a bird. That's why it's known as the canary of the sea. The bowhead whale of the Arctic can sing two different notes at the same time. It can also imitate surrounding sounds, such as the groans of moving ice and the songs of other whales.

When traveling from place to place, bowheads and belugas sing to members of their pods. They whoop, groan, and echolocate to avoid crashing into large chunks of floating ice.

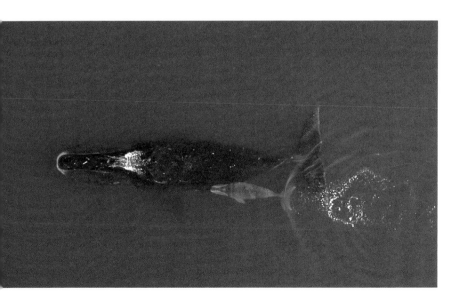

Beluga whales (above) are known as the canaries of the sea. Bowhead whales (left) can imitate the groans of moving ice and the songs of other whales.

A humpback whale comes to the surface for a breath of air.

Another famous singer, related to the bowhead, is the humpback whale. Each winter, herds of humpbacks **migrate,** or move, away from cold northern seas. They head south to warmer waters, where they mate.

As soon as male humpbacks arrive, they start singing to one another or to themselves. At first they sound like singers practicing their scales. High and low notes are mixed with squeaks, chirps, yups, ees, and oos. Then, slowly, different sounds are repeated, like the choruses of human songs, and the music seems to have a beginning, a middle, and an end.

The males sing night and day, and their songs may last for a few minutes or be repeated for several hours. Sometimes it sounds as if the whales are giving a concert. Every now and then they take time out to come to the surface and fill their lungs with air. When they return underwater, each one usually picks up his song at exactly the place where he left off.

All males in the group sing the same song, but each does so in his own special way. In the middle of a crowd of singing males, a female can easily find her mate by the way he is singing.

Humpback whales sing underwater.

After the humpbacks mate and the young are born, the herd moves north again, toward feeding grounds in colder climates. As they travel, members of the pod sometimes swim miles apart, or the young get separated from their mothers. Then they call to one another so that no one will get lost.

The following winter, the whales return to the same breeding ground, and the males again begin their mysterious singing. Theirs is the longest and most complicated song of any underwater musician.

A young humpback whale swims with its mother.

Many sea creatures, like this red lionfish, may become frightened when they hear the loud noises of ships, machinery, and explosions.

Sea creatures around the world make noises in the waters they live in. But people also make noises in the sea. What do the rumblings and roarings of ships and speed boats do to these animals? How do underwater explosions affect the animals? Do these noises make it harder for these animals to find food, attract mates, or fight off predators? Perhaps one day you will discover the answers and help these remarkable creatures continue their music making.

A variety of undersea creatures make different sounds when they search for food, call to their mates, or try to frighten away enemies. Below are a few of these music-making animals and several facts about them.

CREATURE	TYPE	HABITAT	SOUND	HOW SOUND IS MADE	WHEN SOUND IS USED
Croaker	fish	near sea floor	humming	vibrates muscles against swim bladder	during mating season
Grouper	fish	rocks, reefs	moos, grunts	grinds teeth	when threatened
Grunt	fish	coral reefs, sand, mud, grass	grunts	grinds teeth in throat	when calling to one another
Mussel	mollusk	rocks, hulls of ships	popping, snapping	breaks lines that hold it in place	when moving
Pistol shrimp	crustacean	sandy bottoms	pops	rubs two parts of large claw together	when frightened
Dolphin	mammal	offshore waters	clicks	forces air through blowhole	to locate food
Bowhead whale	mammal	near ice in Far North	whoops, groans, purrs	may vibrate tissues in throat	when calling to one another
Humpback whale	mammal	along coastlines, sometimes in open seas	high and low notes, ees, oos, squeaks	may vibrate tissues in throat	during mating season
Walrus	mammal	near frozen ice in Far North	knocks	may snap tongue against roof of mouth	during mating season

Glossary

blowhole: one or two nostrils on the heads of whales and dolphins

crustacean: an animal, such as a lobster, shrimp, or crab, that has a hard, moveable skeleton covering its body

eardrum: a layer of skin inside the ear that vibrates when sound strikes it

echolocation: the use of sound and echoes to find objects

incisors: the four front teeth of some animals

mammal: a warm-blooded animal that has a backbone and hair on some or all of its body. Young are fed milk from their mother's body.

mate: to come together to produce young

migrate: to move from one place to another, usually to reach new feeding grounds

molecule: tiny particles that make up air, water, and many other things

mollusk: a soft-bodied animal that has no backbone and is sometimes covered by a shell

pharyngeal teeth: teeth that line the throats of some fish

pod: a small group of sea mammals, such as whales or dolphins

predator: an animal that hunts and eats other animals

school: a large group of sea creatures feeding or swimming together

species: kinds of animals that are alike in several ways

swim bladder: a gas-filled organ inside some fish that keeps them from sinking when they stop swimming

Index